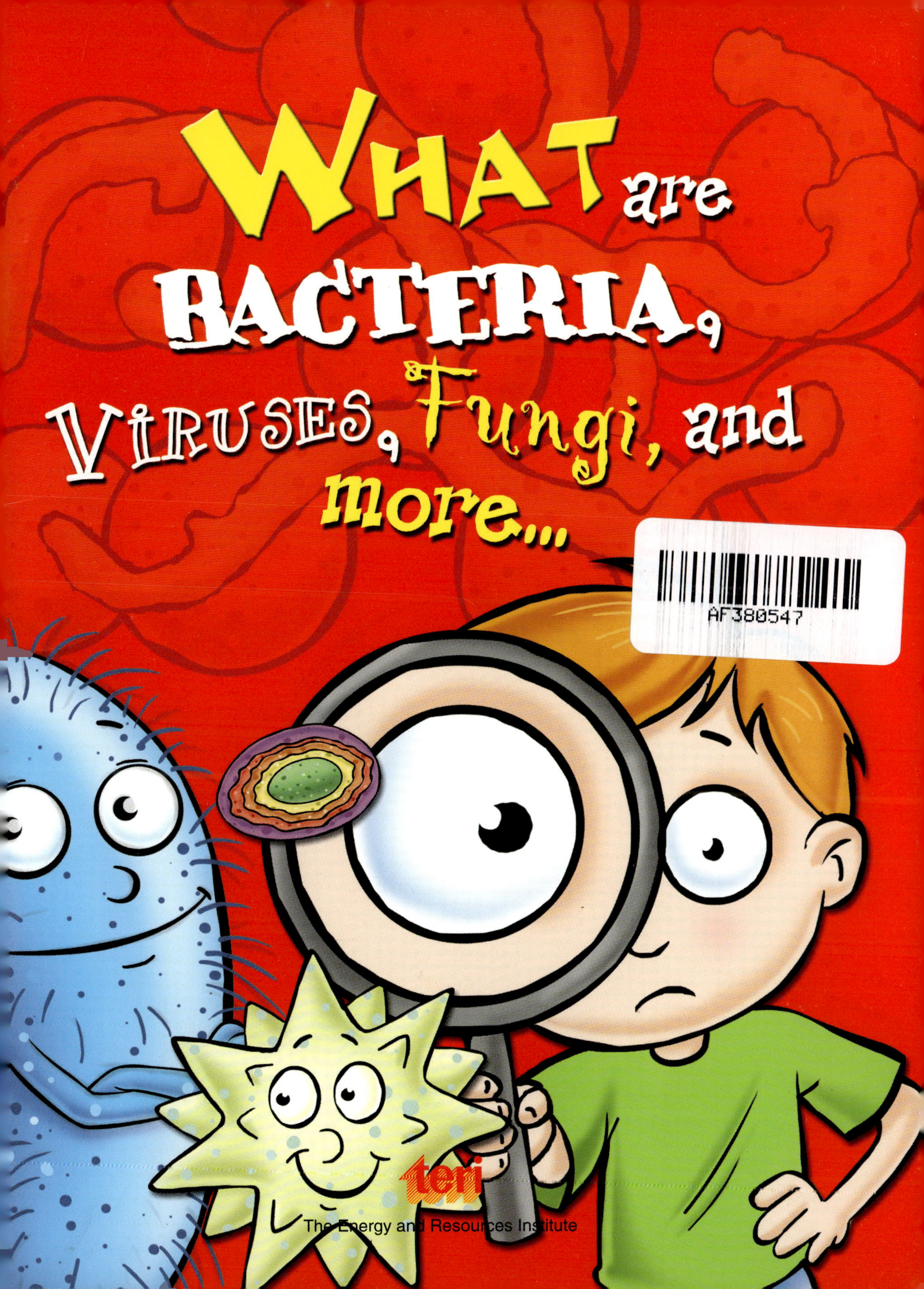

WHAT are
BACTERIA,
VIRUSES, &Fungi, and more...
AF380547
teri
The Energy and Resources Institute

All the Small Creatures

How many times has your mother insisted that you wash your hands before you eat, so that there are no germs on your hands. These germs, called microbes or micro-organisms, are some of the smallest creatures on earth. They are so small that you cannot see them with naked eyes.

A modern microscope can help us unravel the mysteries of even the smallest microbes.

Discovering a whole new world

So, if microbes cannot even be seen, how do we know so much about them? It is thanks to a man named Anton van Leeuwenhoek, who in 1673, used two lenses (lenses are the special glasses that are used in spectacles) to build a microscope. By its use, he discovered a whole new world of tiny microscopic creatures.

The first microscopes looked nothing like they do now!

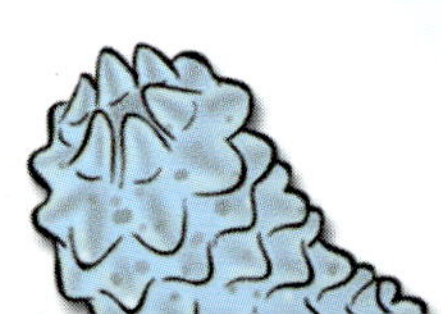

Too many, too small

Microbes are the oldest creatures on our planet and the largest in number. These creatures are generally made up of a single cell. But some microbes like fungi can be made up of many cells, just like your body is made up of billions and billions of cells.

How it all began

Microbes can be prokaryotes or eukaryotes. This depends on the presence or absence of the nucleus. Just like the brain controls all the functions of our body, the nucleus acts as a control system for the cell.

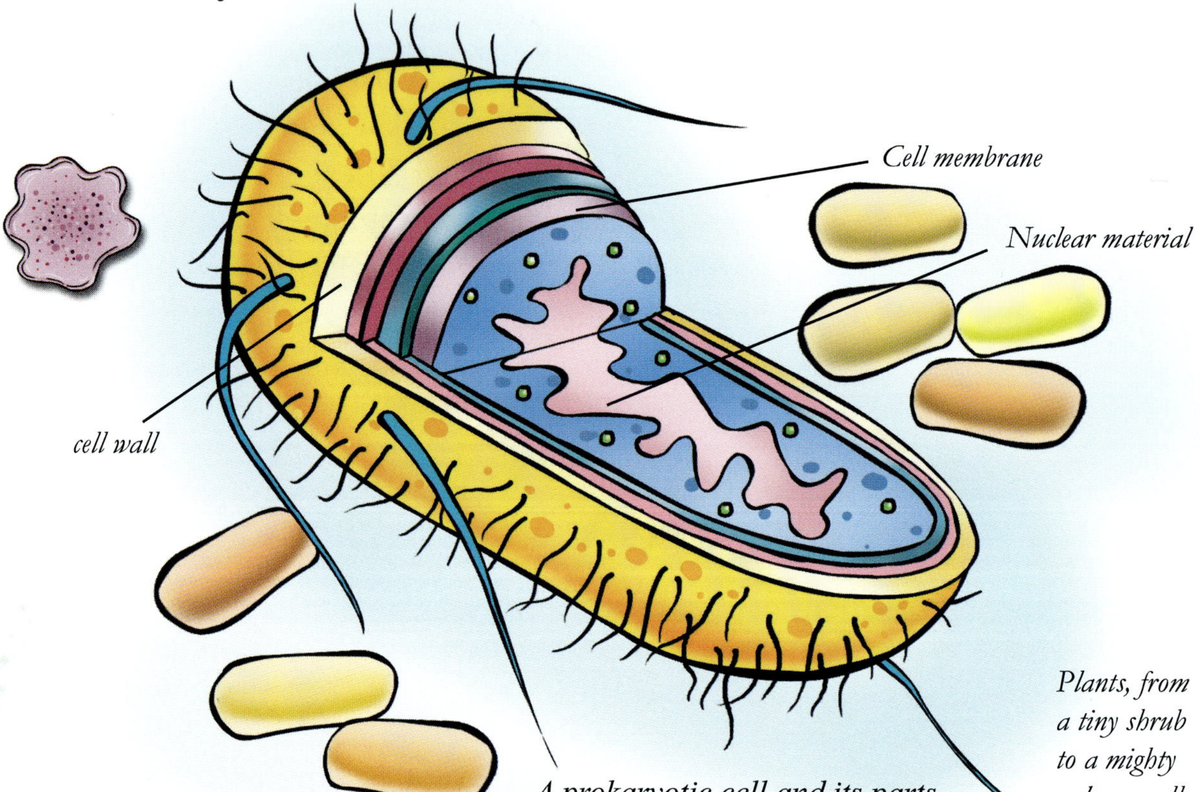

A prokaryotic cell and its parts

Plants, from a tiny shrub to a mighty oak, are all eukaryotes. All their cells have nuclei.

Prokaryotes: surviving without a nucleus

The Greek term "prokaryote" means "before nucleus". Prokaryotes are called so because they do not have a well-defined nucleus. The information on how a cell will function is stored in the deoxyribonucleic acid, or DNA, which floats freely in the cell. Prokaryotic cells are usually round, rod-shaped, or helical.

A eukaryotic cell and its parts

Eukaryotes: ruled by the nucleus

The Greek word "eukaryote" means "true nucleus". Eukaryotes are organisms that have a well-defined nucleus with a marked outer covering. All the information about the cell is stored in the nucleus as DNA. Eukaryotes can be single-celled like amoebae or they may be made up of many cells like the plants and animals we see around us.

Soak up the Sun

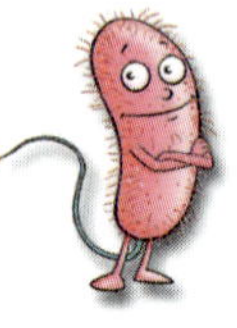

Bacteria have been on the earth since a long time—they appeared even before plants and animals came into being. They are single-celled, prokaryotic, and are present everywhere.

Bacteria thrive in warm climates, where food is abundant, and the weather, pleasant.

Where and what

Bacteria are found in every kind of environment—from the bottom of the deepest oceans to the icy cold Antarctic regions. A single teaspoon of topsoil contains more than a billion bacteria. Some of them can live in temperatures above the boiling point of water and in cold that would freeze your blood. In harsh conditions, they become inactive and form endospores. They remain in this state till conditions become favourable for their survival. They eat almost everything—from sugar and starch to sulphur and even iron.

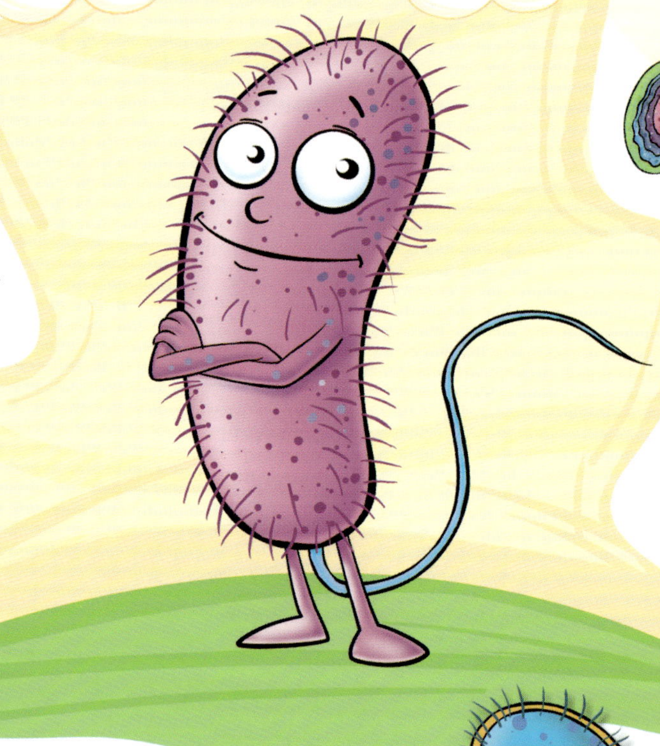

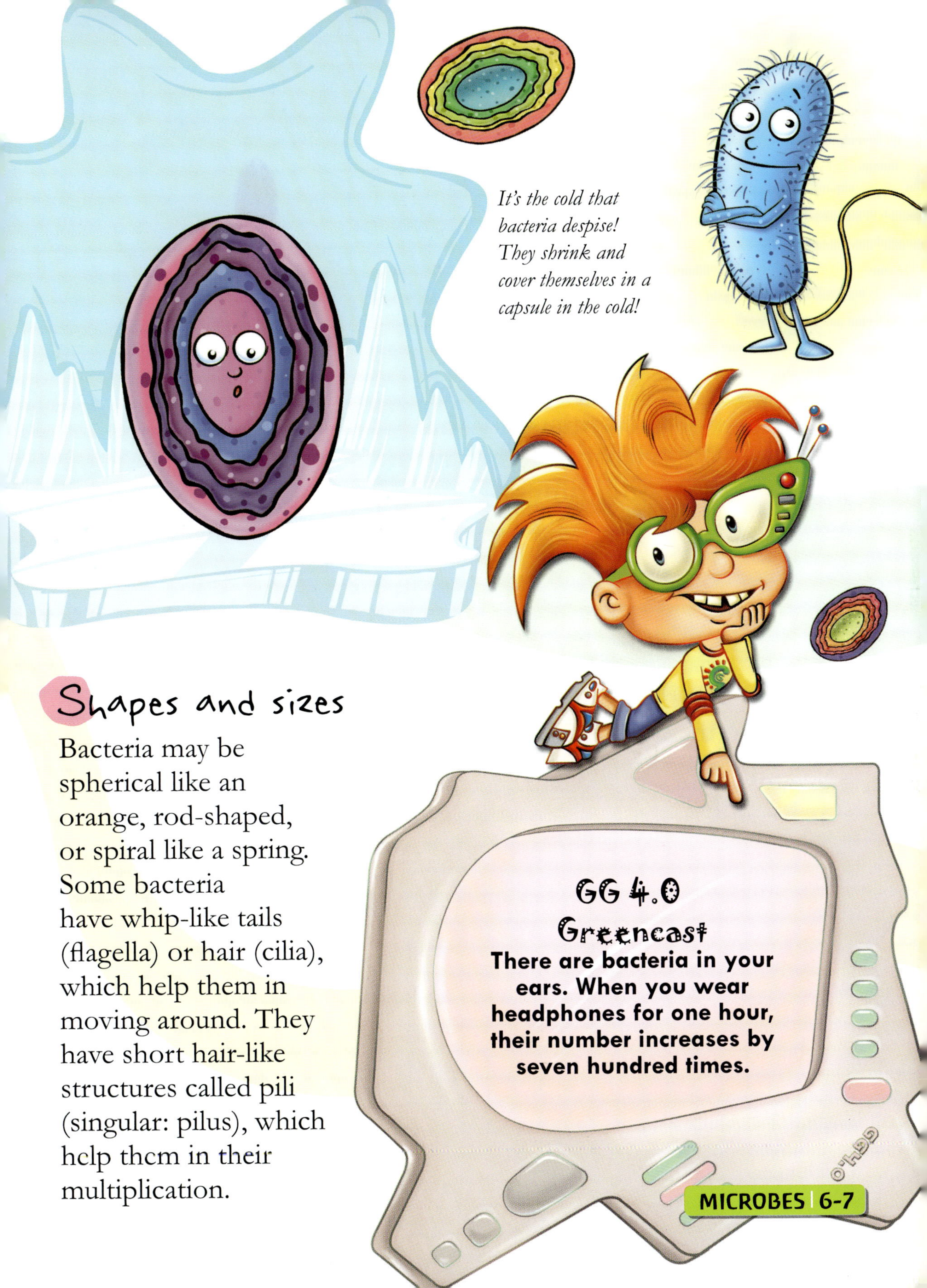

Shapes and sizes

Bacteria may be spherical like an orange, rod-shaped, or spiral like a spring. Some bacteria have whip-like tails (flagella) or hair (cilia), which help them in moving around. They have short hair-like structures called pili (singular: pilus), which hclp thcm in their multiplication.

Making a Killing

Virus in Latin means "poison". Viruses are the smallest of the prokaryotic organisms. Some scientists say that viruses are not even living things.

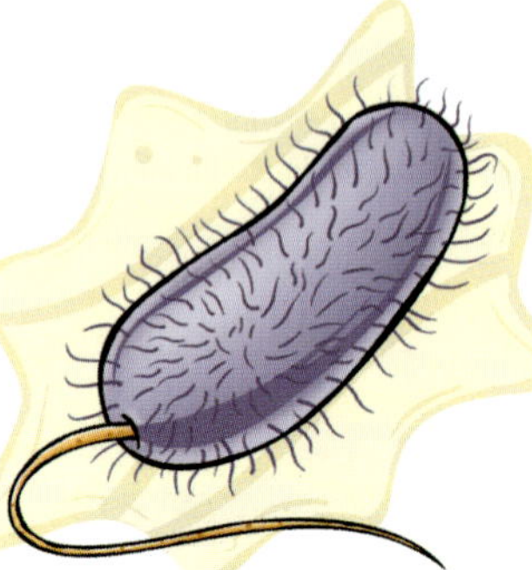

Flagellates have a whip-like flagella on their bodies.

Bacteriophage are viruses that attack bacteria! Some even cure diseases caused by bacteria.

Viruses must eat

All viruses are parasites. They get food and energy from the organism in which they live. Though they reproduce (multiply) like living organisms, they cannot multiply on their own. They need to attack a host organism (another microbe or a plant or an animal) for this purpose.

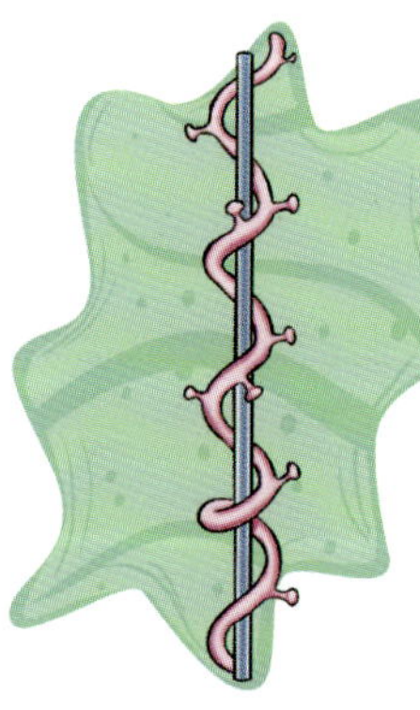

The tobacco mosaic virus is a helical virus.

The shape of a virus

Viruses may be helical (think of a wire wound around a stick) or polyhedric (a structure with many sides) or a mixture of both. The smallest viruses are icosahedral (twenty-faced figure) and the largest are rod shaped. Prions and viroids are even smaller than viruses.

The virus that causes the common cold, rhinovirus, is shaped like an icosahedron.

BEFORE THE REST

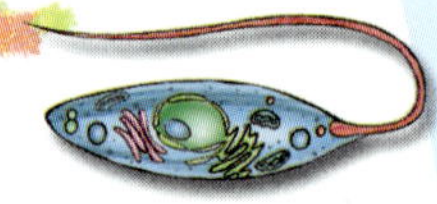

The word "protozoa" is derived from the Greek language. *Proto* means first and *zoion* means animal. Protozoa (singular: protozoan) are eukaryotic and single-celled. They are also known as protists. Although they are made up of only one cell, they breathe, move, and multiply just like multicellular animals.

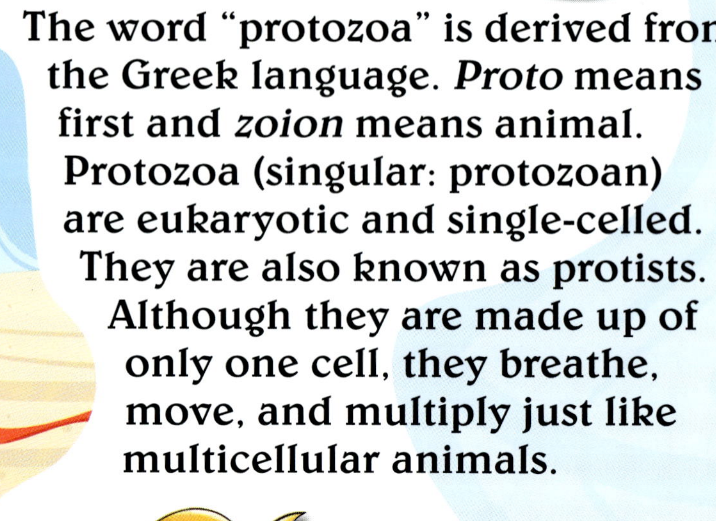

Who came first in the race of living beings? Protozoa!

Where do they live?

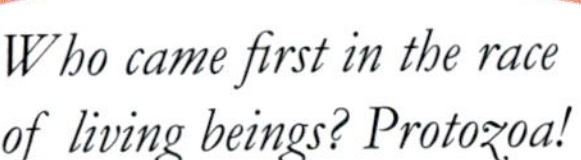

Protozoa live in all kinds of environment— in hot and cold climates, in fresh water and saline water, at high altitudes and below sea level, and even in many plants and animals.

GG 4.0
Greencast
Have you ever seen water in a pond look greenish in colour? This is because of the presence of euglena—a kind of protist.

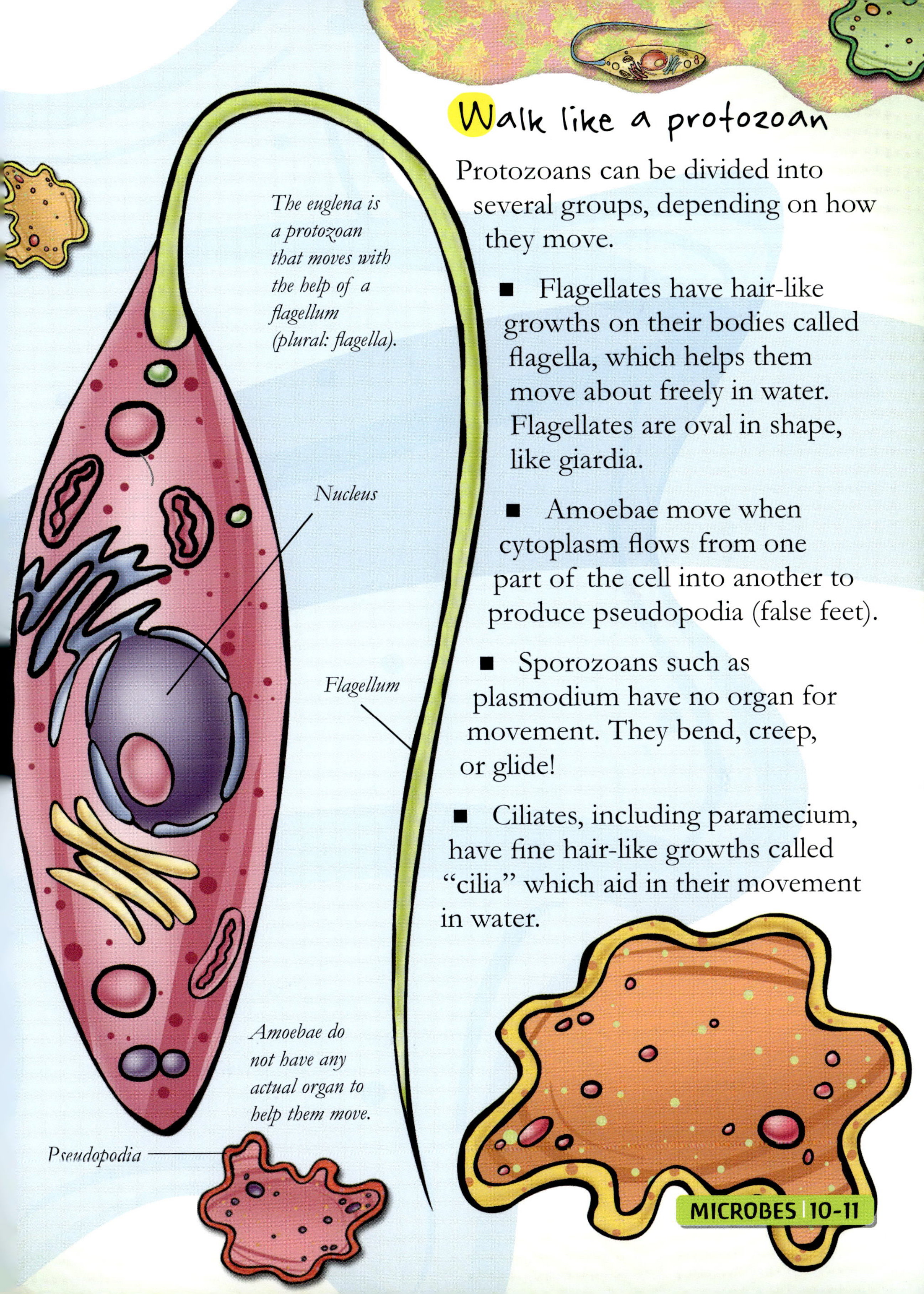

Walk like a protozoan

Protozoans can be divided into several groups, depending on how they move.

- Flagellates have hair-like growths on their bodies called flagella, which helps them move about freely in water. Flagellates are oval in shape, like giardia.

- Amoebae move when cytoplasm flows from one part of the cell into another to produce pseudopodia (false feet).

- Sporozoans such as plasmodium have no organ for movement. They bend, creep, or glide!

- Ciliates, including paramecium, have fine hair-like growths called "cilia" which aid in their movement in water.

ANIMAL? VEGETABLE?

Fungi (singular: fungus, are single-celled or multicellular organisms. Fungi were thought to be plants for many years, before scientists realized that they were very similar to animals too. Now, fungi are classified as a different life form altogether.

Web of life

The "body" of the fungus is a mycelium (thread-like structure). It is made up of a web of tiny hair-like filaments called hyphae. Fungi feed by absorbing nutrients from the plants and animals on which they live. They do this with the help of the mycelium, which is usually hidden in the food source.

Lichens such as the Iceland moss are a combination of both fungi and algae.

How they eat

Certain fungi get attached to the roots of plants, forming a symbiotic relationship. In this relationship, fungi absorb minerals and nutrients from the soil and pass them on to the plants. In turn, they derive their food from the plant. Lichens are made of fungi and algae. Fungi depend on algae for food and in turn, soak up water from the soil for the algae.

Mushrooms are neither plants nor animals. They are fungi!

A fungus is not always bad. Several kinds of mushrooms are used in medicines.

How useful are they?

Some kinds of fungi like some mushrooms can be eaten. Several medicines are made from them. They also break down dcad plants and animals.

Good Microbe, Bad Microbe

Some germs attack our body, get nutrients from the food we eat, and produce substances that are harmful to our health. But microbes don't always cause diseases. Some microbes also perform different functions that are useful for living beings.

What's that you're eating?

Bacteria like rhizobium and nitrosomonas live in the roots of plants and make nitrogen for them. Some bacteria live in the digestive system of animals and help them break down food. They are even used to ferment food such as buttermilk and yogurt.

Setting curd? That's bacteria hard at work.

Clean the world

Microbes are decomposers and break down waste material, thus ridding the earth of all that trash.

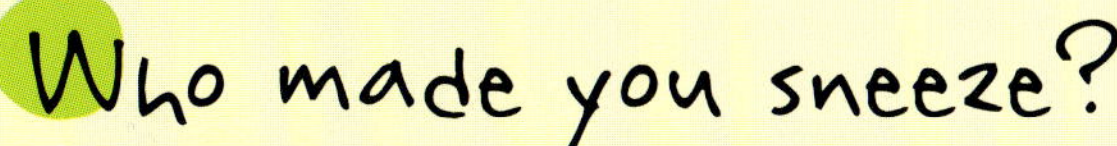

Who made you sneeze?

Viruses cause cold, rabies, AIDS, chickenpox, and many other diseases. Bacteria cause infections and cavities. Fungi cause athlete's foot and ringworm.

Most microbes are spread through the air. They can also spread through sweat, saliva, and blood. There is a simple way to beat these villains. Cover your nose and mouth when you sneeze or cough, and wash your hands frequently with soap and water.

Some bacteria are good for health. They are used to make medicines like penicillin.

When you have a cold and you sneeze, you release germs into the air.

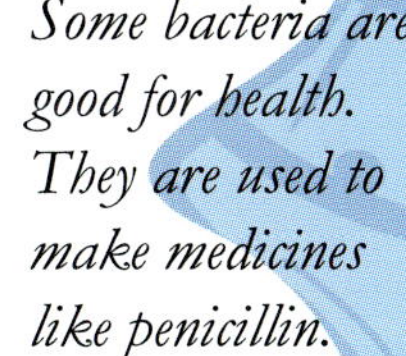

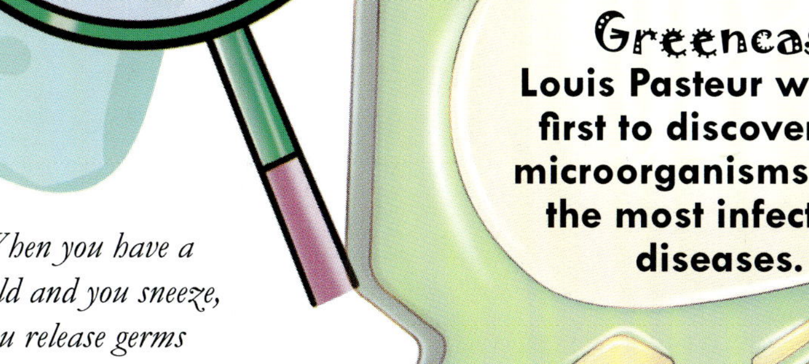

CELL: It is the basic unit of life. A single cell or many cells together make a living body.

CELL MEMBRANE: The outer covering of a cell. It protects the cell and keeps a check on the water, nutrients, and wastes going in and out of the cell.

CELL WALL: The cell wall covers the call membrame. It supports the cell.

CYTOPLASM: It is made up mostly of water and contains all the organs. It is separated from the outside environment by the cell membrane and the cell wall.

DNA: It stands for deoxyribonucleic acid, and contains genes and makes proteins for an organism.

ENDOSPORES: Structures formed by some bacteria to escape the harsh or unfavourable conditions like extreme heat or cold.

MICROSCOPE: An instrument used to see those tiny objects that cannot be seen through naked eyes.

NUCLEUS: The most important organ in a cell. It organizes the working of the cell.

PARASITIC ORGANISMS: Organisms that live on other living organisms and derive their food from them.